Ultimate Origin: ProtoCosmos and HyperCosmos

The Substance of the Cosmos and Universes is Dimension.

Stephen Blaha Ph. D.
Blaha Research

Pingree-Hill Publishing

MMXXII

To Margaret

Some Other Books by Stephen Blaha

All the Megaverse! Starships Exploring the Endless Universes of the Cosmos using the Baryonic Force (Blaha Research, Auburn, NH, 2014)

SuperCivilizations: Civilizations as Superorganisms (McMann-Fisher Publishing, Auburn, NH, 2010)

All the Universe! Faster Than Light Tachyon Quark Starships & Particle Accelerators with the LHC as a Prototype Starship Drive Scientific Edition (Pingree-Hill Publishing, Auburn, NH, 2011).

Unification of God Theory and Unified SuperStandard Model THIRD EDITION (Pingree Hill Publishing, Auburn, NH, 2018).

The Exact QED Calculation of the Fine Structure Constant Implies ALL 4D Universes have the Same Physics/Life Prospects (Pingree Hill Publishing, Auburn, NH, 2019).

Integration of General Relativity and Quantum Theory: Octonion Cosmology, GiFT, Creation/Annihilation Spaces CASe, Reduction of Spaces to a Few Fermions and Symmetries in Fundamental Frames (Pingree Hill Publishing, Auburn, NH, 2021).

Passing Through Nature to Eternity ProtoCosmos, HyperCosmos, Unified SuperStandard Theory (Pingree Hill Publishing, Auburn, NH, 2022).

HyperCosmos Fractionation and Fundamental Reference Frame Based Unification: Particle Inner Space Basis of Parton and Dual Resonance Models (Pingree Hill Publishing, Auburn, NH, 2022).

A New UniDimension ProtoCosmos and SuperString F-Theory Relation to the HyperCosmos (Pingree Hill Publishing, Auburn, NH, 2022).

The Cosmic Panorama: ProtoCosmos, HyperCosmos, Unified SuperStandard Theory (UST) Derivation (Pingree Hill Publishing, Auburn, NH, 2022).

Available on Amazon.com, bn.com Amazon.co.uk and other international web sites as well as at better bookstores.

CONTENTS

FIGURES and TABLES

Introduction

In earlier books such as *The Cosmic Panorama, HyperCosmos Fractionation …,* and *Passing Through Nature to Eternity*, we detailed the chain leading from the ProtoCosmos through the HyperCosmos to the Unified SuperStandard Theory (UST). This book shows that the underpinning of these theories is a primordial substance: Dimension. We see the new ProtoCosmos as a continuous line of dimensions. Each HyperCosmos space is generated by a scaling dimensional force that aggregates the space's dimensions into a finite discrete set. Universes, such as our own universe, build dimensions within a mathematical garb into particles and interactions. A new formalism for unification enables all the dimensions of all the HyperCosmos dimension arrays to be generated in a HyperUnification Space from one dimension element. Thus Creation can be based on one dimension – the ultimate simplicity!

The search for a primordial substance or element has transcended ordinary matter: the fundamental quarks, leptons and so on, to reach a new conceptual foundation built around the primary concept of space—embodied in the concept of dimensions, which, loosely viewed, are the "grains" of which spaces are composed.

This book establishes a new view of the ProtoCosmos as a continuous line of dimensions analogous to the continuous line of angular momentum of a Regge Trajectory. A dimension self-interaction causes a spectrum of spaces to appear containing sets of discrete dimensions: the HyperCosmos ten space spectrum.

Each space uses its dimensions to create universes with each universe containing a set of fundamental fermions, a set of Internal Symmetry groups and gauge fields, a set of Higgs bosons, and so on. The UST is for our universe. Each universe has a Fundamental Reference Frame (FRF) that unifies its Symmetries, General Relativity, and Quantum Field Theory (as generalized by the author's Two-Tier and PseudoQuantum Theory).

The book traces the evolution of the Cosmos from ProtoCosmos through HyperCosmos to UST theories. It closes with the view that the proper goal of Physics and Cosmology is to seek the deepest level.

Dimensions are primitives. They may be counted. But they have no numeric value. They are like the primitives of Euclidean geometry such as point and line. We suggest they are at the deepest level and should be treated as undefined primitives. Everything may be reduced to one primary dimension creating a unified set of spaces and universes. These discoveries not likely to be the result of a yet deeper level of Physics.

The historical trend of Physical primordial elements is now:

Fire-Earth-Air-Water – Atoms – Quarks, Leptons, … – Space – Dimension

1. ProtoCosmos Space

In earlier books we created a ProtoCosmos space that generates the spaces of the HyperCosmos. In this chapter we recap the UniDimension ProtoCosmos model. In chapters 4 and 5 we develop a new formulation of the UniDimension Model in which the continuous UniDimension space becomes a space consisting of one continuous line of dimensions. The HyperCosmos spaces that result are the same. But the reinterpretation of the UniDimension line leads to a new conceptuality of universes in which their primordial element is dimension.

1.1 New UniDimension ProtoCosmos Model

The ProtoCosmos UniDimension Model, as originally conceived,[1] consists of a one dimension "line" having a center of force. (Fig. 1.1) It has a one dimension (relativistic) Dirac equation with a set of space wave eigenfunctions.[2] These space wave functions have an energy spectrum that maps to the HyperCosmos spectrum of spaces.

The UniDimension ProtoCosmos Model appears to be the most minimal formulation for the basis of the HyperCosmos.

1.1.1 UniDimension Force

The UniDimension force in the ProtoCosmos will be initially treated as a one dimension electromagnetic potential. It consists of a one potential $A(x)$ wave function only due to the ProtoCosmos having a one dimension space with no time coordinate. We view this space as the limit of a two dimension space with a time component where the time component is "frozen."

We assume the electromagnetic interaction has a $1/x$ form knowing that a geometric scaling output energy spectrum will result that leads to the HyperCosmos. It is generated from a Lagrangian term:

$$\mathcal{L} = -\tfrac{1}{2}m_A^2 A^2 \tag{1.1}$$

where m_A is a mass-like parameter corresponding to A, which is massless. The dynamic equation for A is[3]

$$m_A^2 A = 0 \tag{1.2}$$

If we restrict x to $0 \le x \le \infty$, and use

$$\int_0^\infty dk\, e^{ikx} = i/(x + i\varepsilon)$$

---

[1] This chapter appeared in Blaha (2022f).
[2] Space wave functions are wave functions that have a set of coordinates of the current universe and a set of coordinates for a "subsidiary" universe. See cjchapter 6 in Blaha (2022c) for details.
[3] Note gauge invariance is absent.

then

$$A = V(x) = 1/(2\pi\, m_A^2) \int_0^\infty dk\, e^{-ikx} = 1/(m_A^2 x) \qquad (1.3)$$

The electromagnetic propagator has the form

$$S(x-y) = <0|T(A(x)A(y))|0> = 1/(2\pi) \int dk\, e^{-ik\cdot(x-y)}/(m_A^2) \qquad (1.4)$$

Note

$$m_A^2\, S\,(x-y) = \delta(x-y) \qquad (1.5)$$

The resulting basic UniDimension Dirac-like equation is:

$$(p + m + V(x))\psi = H\psi = E\psi \qquad (1.6)$$

where m is a fermion mass, V specifies a force center at $x = 0$, and

$$p = \hbar/i\, d/dx \qquad (1.7)$$

$$V(x) = -Z/x \qquad (1.8)$$

for a central charge Z.[4] Below we extend eq. 1.6 to space wave functions incorporating subsidiary universes. Note the Dirac matrix in one dimension is a 1×1 constant that we can take to be unity.

1.1.2 ProtoCosmos Free Space Wave Function

We wish to define a wave function for the UniDimension that embodies both the x line and introduces a subsidiary space, from which the HyperCosmos spaces emerge. We generalize eq. 1.6 to the separable equation

$$[p + m + V(x) + M + i\gamma^\mu\cdot\partial/\partial u^\mu|_{f(E)}]\psi(x, u) = H\psi(x, u) = E\psi(x, u) \qquad (1.9)$$

where the universe u is described by the fermion wave, where x is the ProtoCosmos coordinate and u represents the subsidiary space/universe coordinates. The set of u coordinates consists of $(u_1, u_2, \ldots, u_{f(E)})$ where f(E) is an integer specified as a function of the energy E. We separate eq. 1.9 into

$$\psi(x, u) = \psi_1(x)\psi_2(u) \qquad (1.10)$$
$$[p + m + V(x)]\psi_1(x) = E\psi_1(x) \qquad (1.11)$$

$$[\, i\gamma^\mu\cdot\partial/\partial u^\mu|_{f(E)} + M]\psi_2(u) = 0 \qquad (1.12)$$

---

[4] We absorb an m_A^2 factor into Z.

with the (discrete) E spectrum determined by the ProtoCosmos space x, and f(E), which will be an integer valued function of E, determined for each E separately for each energy value in the spectrum. The energy spectrum of E values will be seen to be geometric. The result will be a set of spaces/universes that correspond to the HyperCosmos set of spaces. The use of f(E) is analogous to the effects of the role of eigenvalues of L^2 (angular momentum) in the Quantum Mechanics Hydrogen atom. The Hydrogen energy E eigenvalues limit the L^2 eigenvalues $l(l + 1)$ in the Hydrogen spectrum.

Case calculated quantum mechanical wave functions for singular potentials in four dimensions. This relativistic four dimension Dirac equation[5]

$$(\beta\gamma^1 p + \beta m + V(x))\psi = H\psi = E\psi \tag{1.13}$$

has the same form as eq. 1.11 above. He calculated the singular potential energy levels based on requiring a fixed phase for wave functions at the origin. This approach guarantees the required orthogonal eigensolutions.

Using Case's method eq. 1.11 leads to an energy spectrum which we may express in the form:

$$E_N \cong m[1 - c_1 c_2^{-2N}] \tag{1.14}$$

where c_1 and c_2 are constants.

We now connect tis energy spectrum to the HyperCosmos spaces spectrum by requiring[6]

$$\begin{aligned} c_1 &= 1/m \\ c_2 &= 2^{22} \end{aligned} \tag{1.15}$$

Then the infinite UniDimension line with the form of Fig. 1.2 has

$$E_N \cong m - d_{dN} \tag{1.16}$$

where

$$d_{dN} = 2^{22-2N}$$

with d_{dN} being the number of components in the dimension array of the Blaha number N space. We note the following HyperCosmos relations, which hold here:[7]

$$\begin{aligned} r &= 18 - 2N = \text{number of space-time dimensions} \\ d_{dN} &= 2^{22-2N} = 2^{r+4} = \text{number of dimension array components} \\ d_{cd} &= 2^{11-N} = \text{column length of the } d_{dN} \text{ array} \end{aligned} \tag{1.17}$$

---

[5] From Blaha (2022c). The UniDimension ProtoCosmos space being one dimensional leads to a Majorana-like fermion wave function $\psi_1(x)$. There is no Electromagetic-like charge.

[6] To obtain consistency with the creation/annihilation operator derivation of the HyperCosmos spaces spectrum.

[7] See Blaha (2022f) for details.

Fig. 1.2 shows the energy spectrum superimposed on the string symbolically. Fig. 1.1 represents the ProtoCosmos string. Below we display the wave functions $\psi_1(x)$ and $\psi_2(u)$. The $\psi_2(u)$ factor is a free fermion wave function that has its form determined by the energy eigenvalues of $\psi_1(x)$.

   We now modify eqs. 1.9 and 1.12 to a form explicitly dependent on the energy.

$$[p + m + V(x) + M + i\gamma^\mu \cdot \partial/\partial u^\mu|_{r\,=\,f(E)}]\psi(x,\,u) = H\psi(x,\,u) = E\psi(x,\,u) \qquad (1.18)$$

$$[\,i\gamma^\mu \cdot \partial/\partial u^\mu|_{r\,=\,f(E)} + M]\psi_2(u) = 0 \qquad (1.19)$$

by making the space-time dimension dependent on energy with[8]

$$f(E) = r = \ln_2[(m - E_N)/16] \qquad (1.20)$$
$$= 18 - 2N$$

by eq. 1.17 where N is the HyperCosmos Blaha number. Thus

$$\gamma^\mu \cdot \partial/\partial u^\mu|_{r\,=\,f(E)} = \sum_{\mu\,=\,1}^{r} \gamma^\mu \cdot \partial/\partial u^\mu \qquad (1.21)$$

Each energy E_N has a different associated space-time dimension r.[9] Fig. 1.1 shows the dependence of the universe space-time dimension on the energy eigenvalue E_N.

   The expression in eq. 1.21 represents an energy dependent term that is analogous to the derivative expression for the angular momentum term L^2 in the Quantum Mechanics hydrogen atom. This L^2 term becomes l (l + 1) where l is restricted by the hydrogen atom energy. Eq. 1.21 may be transformed to a term analogous to L^2 in a deeper more complete UniDimension ProtoCosmos that may eventually emerge.

   The solutions of eq. 1.18 (in the PseudoQuantum form) are:[10]

$$\psi_{iMN\alpha}(\mathbf{x},\,t,\,\mathbf{u},\,t_u) = \psi_{1iN}(x)\,\psi_{2iMN\alpha}(u)$$

where

$$\psi_{iMN\alpha}(\mathbf{x},\,t,\,\mathbf{u},\,t_u) = \psi_{1iN}(x)\,\Sigma_s\int d^{r-1}q\,\mathfrak{N}(q)[u(q,\,s)b_{iMN\alpha}(q,s)e^{-iq\cdot u} + v(q,\,s)d_{iMN\alpha}(q,s)^\dagger e^{iq\cdot u}]$$
$$(1.22)$$

$$\psi_{2iMN\alpha}(\mathbf{u},\,t_u) = \Sigma_s\int d^{r-1}q\,\mathfrak{N}(q)[u(q,\,s)b_{iMN\alpha}(q,s)e^{-iq\cdot u} + v(q,\,s)d_{iMN\alpha}(q,s)^\dagger e^{iq\cdot u}] \quad (1.22a)$$

---

[8] The factor of 16 figures prominently in the Fundamental Reference Frame discussions as setting the size of the basic unit of dimensions. See chapters 9 and 10, and particularly section 10.3, of Blaha (2022c) for details.

[9] Note that the Hydrogen atom Schrödinger equation has an energy dependent range of angular momentums—an analogous feature.

[10] We separate the differential equation to obtain the solution in eq. 2.22. In chapter 6 of Blaha (2022c) we form a combined solution for the universe part and the internal universe parts. Physically, the solution eq. 2.22 reflects the separation of the one dimension ProtoCosmos from the subsidiary universes.

for $i = 1, 2$ where $\mathfrak{N}$ is a normalization constant, where m is the mass in the ProtoCosmos space and M is the mass-energy in the universe space of Blaha number N, r is the number of universe space-time dimensions, and α ranges from 1 through d_{dN}. We assume $q^2 = M^2$ in the universe space. Note $u = (u^0, \mathbf{u})$, and $t_u = u^0$.

The solution eq. 1.22 gives a space fermion of the type described in Blaha (2022c). The range of the symmetry index α from 1 through d_{dN} gives a dimension array of d_{dN} elements that can be subdivided into a d_{cd} by d_{cd} square array of the type seen in the HyperCosmos spaces spectrum. Universe states can be created from sums over α of the wave functions $\psi_{iMN\alpha}(\mathbf{x}, t, \mathbf{u}, t_u)$.

1.2 Universe States

The d_{dN} wave functions may be used to construct universe states, which initially before interactions takes hold have forms of the type:

$$|U> = \prod_{\alpha} b_{iMN\alpha}(q,s)^{\dagger}|0> \qquad (1.23)$$

After interactions"begin" at the Big Bang point $|U>$ acquires a complex form. A universe thus has a complete set of internal symmetry components.

1.3 The Spaces of the UniDimension ProtoCosmos

This ProtoCosmos Model has two relevant types of space-times. First there is the one dimension ProtoCosmos "space-time." There are also the space-times of each of the model's eigensolutions. For each energy E_N there is the r space-time dimension wave function of eq. 1.22. There is also an internal r space-time dimension set of coordinates within the dimension array d_{dN}. The number of these internal space-time coordinates is the value $r = 18 - 2N$. Also the f(E) condition sets r for the wave function.

The internal number of space-time dimensions always equals the number of wave function space-time dimensions due to the f(E) condition. Therefore we are allowed to identify the wave function space-time with the internal space-time of each eigensolution. For each eigensolution there is one dimension ProtoCosmos space and one r dimension internal space-time. The f(E) condition causes the wave function and internal space-times of the model to be the same space-time.

Each universe of each HyperCosmos space has a wave function of the form of eq. 1.22a that exists *within* the universe. Consequently it is possible to second quantize a universe consistently within it. This may be viewed as remedying previous attempts to quantize a universe from "outside" the universe.

The eigenstates of the UniDimension ProtoCosmos correspond to the ten universe spaces of the HyperCosmos as well as the fractional spaces of the Limos (Figs. 1.2 and 2.1). The positive energy ProtoCosmos spectrum is limited by 0 (an accumulation point) and m. The limits of the HyperCosmos spectrum follow.

Figure 1.1. A segment of the one dimension space embodied in a string of infinite length extending from $-\infty$ to ∞. The center of force is indicated by the filled circle.

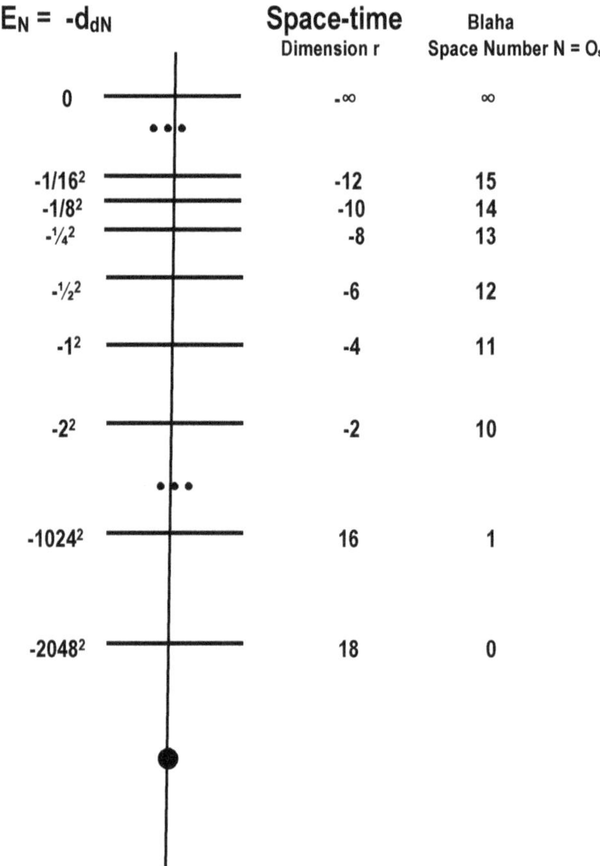

Figure 1.2. The energy level $E_N = -d_{dN}$ aggregates of the UniDimension ProtoCosmos Model (not drawn to scale). The various quantities match the HyperCosmos spaces spectrum in Fig. 2.1. The energy levels are symbolically depicted above. Negative N levels are not shown.

2. The HyperCosmos Spaces

This chapter summarizes the HyperCosmos spectrum of spaces. The spaces of the HyperCosmos with space-time dimensions greater than or equal to zero may support universes including our universe and the Megaverse (Multiverse). The spaces with space-time dimensions less than zero do not support universes. They are the Limos sector of the HyperCosmos space spectrum. Their role is to support the fractionation of fundamental particles as described in Blaha (2022d).

The HyperCosmos spaces have associated universe instances. The Limos sector may be used to describe the untra-small: the interior of fundamental particles. See Blaha (2022d).

2.1 Evidence for the HyperCosmos

There is evidence for the ProtoCosmos and the HyperCosmos. It is not direct evidence since we are bound by our universe and its elapsed lifetime. It is evidence that indirectly supports their existence and their structure.

1. Direct Generation of the HyperCosmos spectrum from a UniDimension ProtoCosmos.

2. Relic Structure in Internal Symmetries and the Fundamental Fermion Spectrum. The Standard Model internal symmetry $SU(2) \otimes U(1) \otimes SU(3)$ is shown to appear in HyperCosmos and QUeST as a natural separation not requiring symmetry breaking Higgs fields. QUeST introduces additional symmetries that add to those of the Standard Model in a direct way—again without Higgs bosons. Higgs breaking is restricted to ElectroWeak-like sectors. See chapter 11 of Blaha (2022f).

3. *The separation of fermions and symmetries into superluminal and subluminal parts in the Fundamental Reference Frame leads to the fundamental fermion spectrum, to the set of internal symmetries of the Standard Model, and to the understanding of the relation of the Higgs Mechanism and Strong Interaction Confinement (the Higgs-Confinement Dichotomy). See chapter 9 of Blaha (2022f).*

4. The Dual Resonance Model of hadron scattering is *directly* based on the HyperCosmos using fractionated hadrons. This approach is an alternative to String Theories. This approach supports hadronic Veneziano amplitudes. See chapter 7 of Blaha (2022d).

5. The Parton Model of Deep Inelastic Lepton-Nucleon Scattering may be viewed as based on fractionated particles. One finds that it "explains" quasi-free partons, the apparent lack of strong interactions between partons thus giving parton freedom, and scaling in x, as well as other features. See chapter 6 of Blaha (2022d).

6. The fractionated particle formalism has lattice theory as a restricted special case. See chapter 12 of Blaha (2022c).

7. The HyperCosmos leads to a deep unification based on its Fundamental Reference Frame formalism. For example, in the case of our universe, there is a unified 12 space-time dimension formalism that generates the unified General Relativity and Internal Symmetries of our universe. See chapter 5 of Blaha (2022d).

The above evidence – particularly the form of the internal symmetries in The Standard Model and UST as shown in chapters 9 and 11 of Blaha (2022f) – leads to the serious consideration of the HyperCosmos.

2.2 Structure of the HyperCosmos Spaces Spectrum

The structure of the HyperCosmos spectrum of spaces follows from a number of considerations:

1. There is a spectrum of spaces in the Cosmos from which universes are instantiated (created).
2. Physically the least space-time dimension of a space is zero.
3. The number of space-time dimensions differs by two between adjacent spaces of the spectrum of spaces.[11]

    Justification:

    In a space-time of even dimension r the least spin fermion has $2^{r/2-1}$ spin states. In a space-time of odd dimension $r' = r + 1$ the least spin fermion *also* has $2^{r/2-1}$ spin values.

    The total number dimensions in a space d_{dN} satisfies[12]

$$d_{dN} = \text{Number of creation/annihilation operators} = 2^{r+4}$$
$$= 2^6 \times (\text{number of spin states})^2 \tag{2.1}$$

    where (number of spin states) $= 2^{r/2-1}$.

    Thus the total number of dimensions d_{dN} in an even r dimension space-time and in an $r' = r +1$ dimension space-time d_{dN}' are equal $d_{dN} =$

---

d_{dN}'. As a result there is a possible redundancy in d_{dN}. This redundancy is removed by assuming r is even since r = 4 for our universe. The spectrum spacing in r is two dimensions.

4. We assume there are ten spaces in the Cosmos spectrum consistent with the UniDimension ProtoCosmos spectrum.

5. Our universe has four space-time dimensions. Thus it is the third space above the space with a zero space-time dimension. See Fig. 2.1.

6. The d_{dN} dimension array of each space is determined by enumerating the creation/annihilation operators of the lowest spin fermion wave functions expressed in the PseudoQuantum formulation for each space-time dimension r. See chapter 6 of Blaha (2022f) for a detailed derivation giving d_{dN} = the number of creation/annihilation operators = $2^{r+4} = 2^{22-2N}$. For example, for r = 4 (Blaha number N = 7) we find d_{dN} = 256 for our universe. See Fig. 2.1.

Thus the structure of the HyperCosmos spectrum of spaces is determined by the nature of fermion PseudoQuantum wave function form, the requirement of a space of which our universe is an instance, and the assumption of a ten space spectrum. Blaha (2022f) contains most of the details.

THE HYPERCOSMOS SPACES SPECTRUM

Blaha Space Number	Cayley-Dickson Number	Cayley Number	Dimension Array	Dimension Array column length	Space-time-Dimension Size	CASe Group $su(2^{r/2},2^{r/2})$	Fermion Spin	Unification Space	
$N = o_s$	n	d_c	d_{cd}	d_{dN}	r	CASe	s	r'	d_{dN}'
0	10	1024	2048	2048^2	18	su(512,512)	255/2	40	2048^4
1	9	512	1024	1024^2	16	su(256,256)	127/2	36	1024^4
2	8	256	512	512^2	14	su(128,128)	63/2	32	512^4
3	7	128	256	256^2	12	su(64,64)	31/2	26	256^4
4	6	64	128	128^2	10	su(32,32)	15/2	22	128^4
5	5	32	64	64^2	8	su(16,16)	7/2	18	64^4
6	4	16	32	32^2	6	su(8,8)	3/2	14	32^4
7	**3**	**8**	**16**	$\mathbf{16^2}$	**4**	**su(4,4)**	**½**	**12**	$\mathbf{16^4}$
8	2	4	8	8^2	2	su(2,2)	0	8	8^4
9	1	2	4	4^2	0	su(1,1)	- ¼	4	4^4

Limos:

10	0	1	2	2^2	-2	U(1)	3/8		0
11	-1	½	1	1^2	-4	U(½)	7/17		-4
12	-2	¼	½	$½^2$	-6				
13	-3	1/8	¼	$¼^2$	-8				
14	-4	1/16	1/8	$1/8^2$	-10				

Figure 2.1. The HyperCosmos space spectrum. Spaces with non-negative space-times may have universe instances. Spaces with negative space-times form the Limos subset of spaces, which are displayed in the shaded area. The d_{cd} column of Limos with $d_{cd} < 1$ specifies the fractionation that may be applied to particles to produce partons, hadron "parts", and so on.

3. Transformations for Reduction to a FRF and the HyperUnification of Space-time and Internal Symmetries

This chapter provides a formalism for the reduction of a dimension array d_{dN} to a transformation from a small set of dimensions in a Fundamental Reference Frame (FRF. The formalism enables the unification of General Relativity and Internal Symmetries for both one space and jointly for all ten HyperCosmos spaces.

3.1 Fundamental Reference Frame (FRF) Transformations

In earlier books, such as in Blaha (2022e), we showed that each of the ten HyperCosmos spaces has a Unification space of a similar type which unites Internal Symmetries with General Relativity within the framework of GiFT.[13]

The basis for this unification is General Relativistic transformations that mix fermion creation and annihilation operators. For example, Blaha (2022f) has the transformation example

$$b_{1\beta s} = \Sigma_{\alpha,x,s} \, (g_\beta, f_\alpha) \, u^\dagger_{\beta s} u_{\alpha s} \, (c_{11s}b_{1\alpha s} + c_{12s}b_{2\alpha s} + C_{11s}b^\dagger_{1\alpha s} + C_{12}b^\dagger_{2\alpha s} + \\ + c'_{11s}d_{1\alpha s} + c'_{12s}d_{2\alpha s} + C'_{11s}d^\dagger_{1\alpha} + C'_{12s}d^\dagger_{2\alpha})$$

which transforms the b annihilation operator into a combination of creation and annihilation and annihilation operators (with similar expressions for the other transformed b and d operators.) Chapter 6 of Blaha (2022f) provides additional detail.

The role of the General Relativistic transformations may be expanded by including Internal Symmetry transformations. Then we can define transformations that transform a small subset of dimensions to the full set of dimensions within a dimension array d_{dN}. We begin by making the d_{dN} dimension array into a vector that is subject to transformations having the form of a $d_{dN} \times d_{dN}$ square array which we denote T. The space of the d_{dN} vector and T must have a dimension array d_{dN}' with d_{dN}^2 entries as shown in Fig. 2.1. Treating d_{dN}' as a HyperCosmos space dimension array and using its relation (chapter 8 of Blaha (2022f)) to its space-time dimensions:

$$r = 18 - 2N \qquad\qquad (8.2)$$
$$d_{dN} = 2^{22-2N} = 2^{r+4}$$
$$d_{dN}' = d_{dN}^2$$

where r is the dimension of the space-time, and d_{dN} is the number of elements in the dimension array for Blaha space N. These equations imply

---

[13] GiFT (<u>G</u>eneral<u>i</u>zed <u>F</u>ield <u>T</u>heory) is compssed of Two Tier and PseudoQuantum Theory within a Quantum framework. See Blaha (2021i) and (2021j) for the original description.

$$r = \log_2 (d_{dN}/16) \tag{8.3}$$

We find the space-time dimension of the space with the $d_{dN}' = d_{dN}{}^2$ dimension array is

$$r' = \log_2(d_{dN}{}^2/16) \tag{3.1}$$

The values of r' for each HyperCosmos space are tabulated in Fig. 2.1. For the space of our universe we find r' = 12.

Thus the transformations have the form of General Relativistic transformations in the space with space-time dimension r'. These transformations provide a unification within the GiFT Quantum Field Theory[14] framework of General Relativity and Internal Symmetries. See Blaha (2022f).

3.1.1 Fundamental Reference Frames

The transformations enable a mapping between a conventional space-time reference frame and a type of frame that we called a Fundamental Reference Frame (FRF. The dimension array of a frame has d_{dN} entries which we represent as a vector. A transformation T may be viewed as transforming a FRF vector of d_{dN} components, of which there are $d_{dN}{}^{1/2}$ non-zero with the remaining entries being zero.

As shown in Blaha (2022f), and earlier books by the author, a transformation T creates $d_{dN}{}^{1/2}$ *replicas* of the $d_{dN}{}^{1/2}$ FRF non-zero entries, thus generating the d_{dN} dimension array.[15] For our universe the dimension array d_{dN} = 256, and its FRF has 16 non-zero entries, with 16 replicates generated by a transformation.

The dimensions define the form of the fundamental fermion spectrum, the dimensions of the irreducible representations of the internal symmetry groups, the gauge vector boson spectrum, and the Higgs boson spectrum. See Blaha (2022f) for details.

3.2 Content of the FRFs

In earlier work in 2020 we noted that each HyperCosmos space had a dimension array that consisted of a set of replicates that separated the fundamental fermions into a set of replicates and separated internal symmetries into a set of replicates. In the case of the space for our universe there were 16 replicates of sets of 16 fermions and 16 replicates of 16 irreducible symmetry group real dimensions.

This pattern of replication held for all HyperCosmos groups. The pattern originates in the Cayley numbers that appear in the ProtoCosmos geometric scaling energy spectrum. Each resulting HyperCosmos space has a dimension array with four times the size of the next lower HyperCosmos dimension array due to the Cayley numbers used to specify dimension arrays. Thus each HyperCosmos dimension array may be viewed as containing four replicates of the next smaller HyperCosmos

---

[14] We note that Physically Quantum Mechanics is a corollary to Quantum Field Theory. See Heitller's book for details.

[15] The FRF vector has $d_{dN}{}^{1/2}$ non-zero entries and $d_{dN} - d_{dN}{}^{1/2}$ zero entries. See chapter 9 of Blaha (2022c) for details.

dimension array. Fir a discussion of these dimension array splittings into replicates see Blaha (2021b) and (2022f).

These observations led us to define the contents of the FRF vectors as having 2^{r+4} elements, of which $2^{r/2+2}$ elements were non-zero dimensions. In the case of fundamental fermions the FRF had $2^{r/2+2}$ fundamental fermions. In the case of irreducible symmetry dimensions there are $2^{r/2+2}$ dimensions. In each case these items were embedded in a 2^{r+4} element FRF vector where $2^{r+4} - 2^{r/2+2}$ elements were zeroes.

Upon General Relativistic transformation to a "normal" reference frame the $2^{r/2+2}$ FRF elements are mapped to $2^{r/2+2}$ replicates yielding a $d_{dN} = 2^{r+4}$ dimension array.

The contents of FRFs can be based on the splittings into replicates as above. However the contents of the FRFs can be different with consequent changes in the T transformations. Later we show that FRFs may be defined with only one non-zero element.

3.3 Unification for the Space of Our Universe

The 256 dimension array of our universe (in the Unified SuperStandard Theory) may be viewed as generated by a Quantum General Relativistic transformation in a 12 space-time dimension unification space. Unification is achieved through the effects of the Quantum General Relativistic transformation of the 12 space-time dimension space. See section 10.2 of Blaha (2022f) for more details.

3.4 Unification for All HyperCosmos Universes

Unification in the other HyperCosmos spaces uses the same procedure. Each space has an associated unification space. The higher space and the "lower" space have space-time dimensions related by:[16]

$$r^{\varepsilon} = 2r + 4 \qquad\qquad (3.2)$$

based on the relation of the r dimension transformation and the r' dimension transformation. To obtain the dimension array for the r coordinate space-time we can use the r^{ε} coordinate space-time. The r' space-time transformation is a $2^{r'/2+2} \times 2^{r'/2+2} = 2^{r+4} \times 2^{r+4}$ array that generates a $2^{r'/2+2} = 2^{r+4}$ coordinates vector from an FRF vector. This generated vector maps to the r space-time dimension array d_{dN}.[17] *The transformations are General Relativistic transformations in the higher space. These transformations embody both the General Relativistic space-time and Internal Symmetries of the lower space.*

3.5 Combined Unification for All Ten HyperCosmos Spaces

In this section we develop a combined General Relativistic transformation for all ten HyperCosmos spaces. From Fig. 2.1 we see that each HyperCosmos space has a specific unification space dimension vector with d_{dN} components.

A unification space transformation can be represented as

---

[16] See Blaha (2022d) and02022e) for more detail.
[17] The array is divided into a square d_{dN} array.

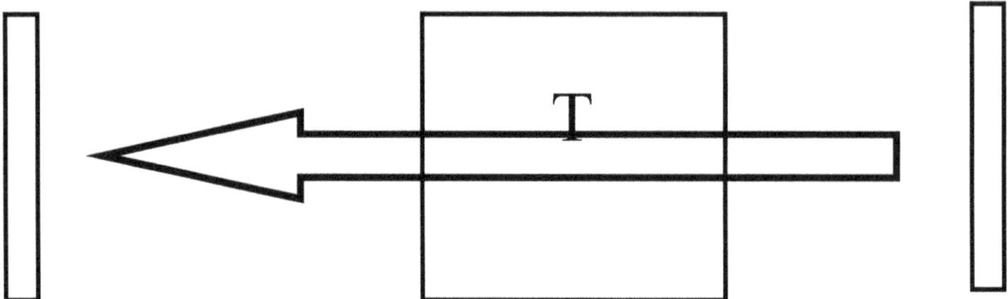

with each vector having d_{dN} components.

3.5.1 HyperUnification

We now consider combined unification of all or some of the HyperCosmos spaces. Each space has a dimension array with d_{dN} elements. This array is reexpressed as a vector v_N with d_{dN} components residing in the space's unification space with space=time dimension r' given by eq. 3.1.

Now consider the combination of all ten spaces where the number of elements is

$$v_S = \sum_{N=0}^{9} v_N \tag{3.3}$$

The combination of the ten vectors of the ten unification spaces vectors is a vector of v_S elements in a larger space with dimension array d_{dS}.

$$d_{dS} = v_S^2 \tag{3.4}$$
$$= 5{,}592{,}400^2$$

The larger space plays the role of the unification space for all ten HyperCosmos spaces. We call it the *HyperUnification Space*. Each of the ten HyperCosmos unification space subvectors in v_S infividually has d_{dN} components. Correspondingly, each subvector has a $d_{dN} \times d_{dN}$ square array along the diagonal of a combined transformation which we call a *HyperUnification Transformation*.

Using eq. 3.1 we find the space-time associated with a dimension array with d_{dS} elements is

$$r_S = \log_2(d_{dS}^2/16) \tag{3.5}$$
$$= 2 \log_2(5{,}592{,}400) - 4 = 40.83$$

We will define a HyperUnification Space with $r_S = 42$ space-time dimensions. The General Relativistic transformations of the $r_S = 42$ HyperUnification Space includes the General Relativistic transformations of any or all, HyperCosmos unification spaces along its "diagonal" where blocks correspond to individual unification space General Relativistic transformations. See Fig. 3.1. It also includes much more in the form of arbitrary, not necessarily block diagonal, transformations that mix various HyperCosmos unification spaces.

We can define a set of HyperUnification Space General Relativistic transformations that transform combinations of all 10 unification spaces. This set of transformations is a subset of the HyperUnification Space General Relativistic transformations

The FRF vector for HyperUnification Space has 2^{23} = 8,388,608 components. Comparing it to the size of d_{dS} = 5,592,400-vector we find the General Relativistic transformations of the 10 HyperCosmos unification spaces are a subset of the 42 space-time dimension HyperUnification Space transformations. The dimension excess is 2,796,208 dimensions.

We will use HyperUnification Space transformations in the next chapter to reduce the set of all HyperCosmos spaces'dimensions to one universal, primordial dimension.

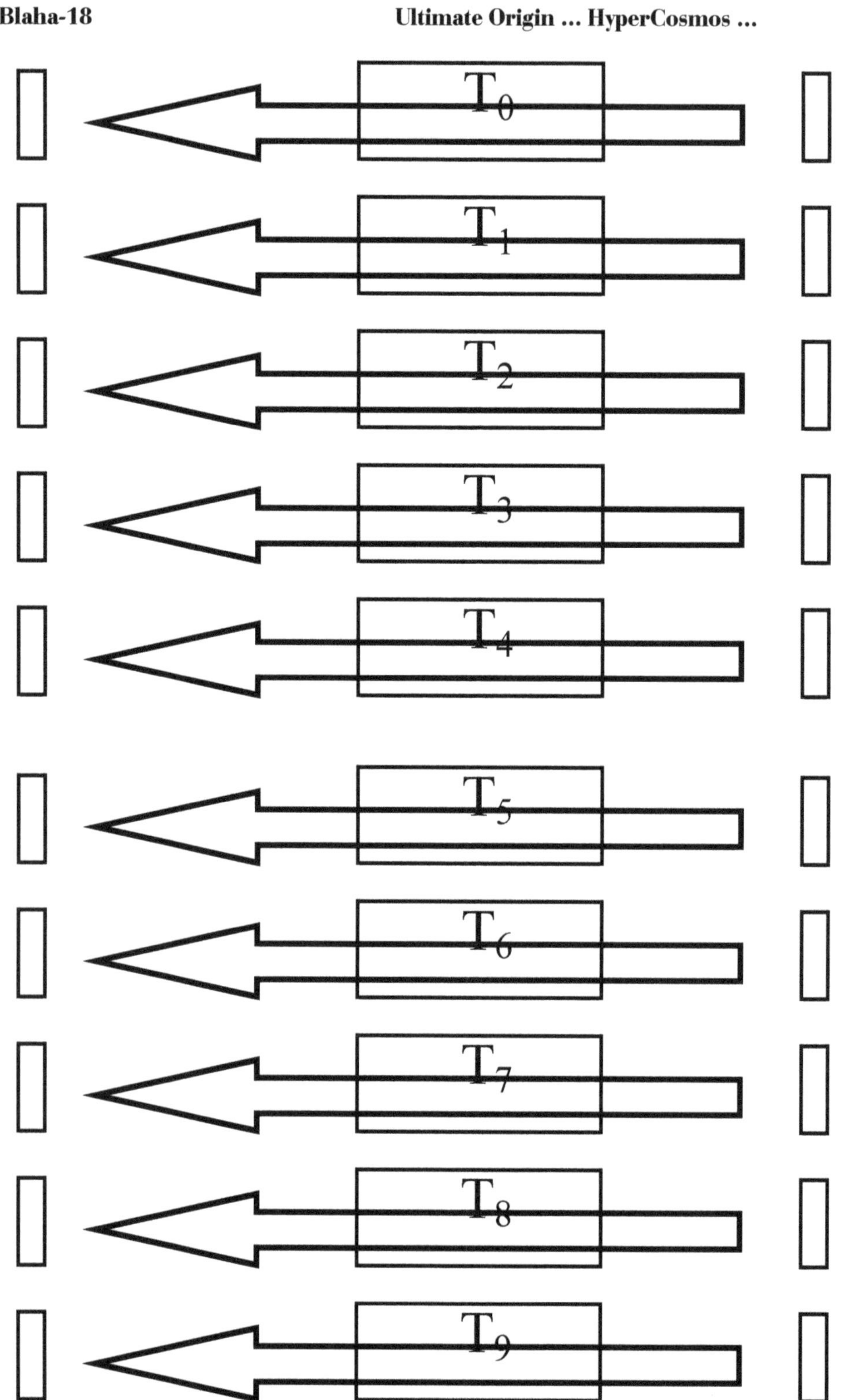

Figure 3.1. HyperCosmos FRF transformations in the HyperUnification Space.

4. Derivation of the HyperCosmos from One Dimension

This chapter describes a reduction of the dimension array of a HyperCosmos space to a unification space General Relativistic transformation from a FRF containing only one dimension. As a result one can envision a space with a large dimension array with d_{dN} entries as a transformation from a one dimension entity. Thus all the fundamental fermions of a space can be generated from one generic fermion. Also all the symmetry group irreducible representation dimensions are from one dimension. All the Higgs bosons from one generic boson. And so on.

Taken together with the "reinterpretation" of the UniDimension ProtoCosmos Model in chapter 5 as being based on a continuous line of dimensions constrained by a self-force, the concept of Dimension as a fundamental substance (element) emerges.

4.1 Transformation from a One Dimension FRF

We demonstrate this possibility with a simple example. Assuming the FRF has one non-zero dimension (element) in row 1 of the FRF vector with all other components zero we define the vector with d_{dN} components as

$$V = (a, 0, 0, \ldots) \qquad (4.1)$$

and a prototype General Relativistic transformation

$$[T]_{ij} = b_i \, \delta_{i1} + (1 - \delta_{1i})(1 - \delta_{j1})c_{ij} \qquad (4.2)$$

where b_i and the c_{ij} are numeric. T has entries b_i in column 1 and a matrix c_{ij} in the rows and columns for $i, j = 2, 3, \ldots$. A full set of d_{dN} array dimensions are generated from one FRF dimension (element).

4.2 Generation of the HyperCosmos Dimensions from One Dimension

The HyperUnification Space defined in chapter 3 can also support the generation of the *entire* set of HyperCosmos dimensions from one dimension in its FRF vector in the HyperUnification Space. Constructing a V with one non-zero dimension as in eq. 4.1 we define a transformation T that maps it to a totally filled non-zero set of d_{dN} arrays for all the ten HyperCosmos spaces. T has the form of eq. 4.2 in the HyperUnification Space.

Thus we can generate the complete set of HyperCosmos dimension arrays from a one dimension element.

5. An Ultimate Origin of the ProtoCosmos

The successful derivation of the dimension arrays for all ten HyperCosmos spaces from a one dimension element in the HyperUnification Space and the derivation of the HyperCosmos spectrum from a Dirac equation in the ProtoCosmos raises a number of questions. The primary question is the basis for the origin of the dimensions of HyperCosmos spaces.

In this chapter we reinterpret the one dimension coordinate space of the ProtoCosmos to be a one dimension continuous line of dimensions. The HyperCosmos spaces then emerge as discrete aggregates of dimensions, which in turn can be reduced to one element in the HyperUnification Space FRF. With this backdrop of features we are led to a derivation of Everything from one substance – dimension – in chapter 6..

5.1 A New Interpretation

The basis for the reinterpretation as a line of dimensions results from a consideration of the Regge Trajectory spectrum of HyperCosmos spaces appearing in Fig. 5.1.

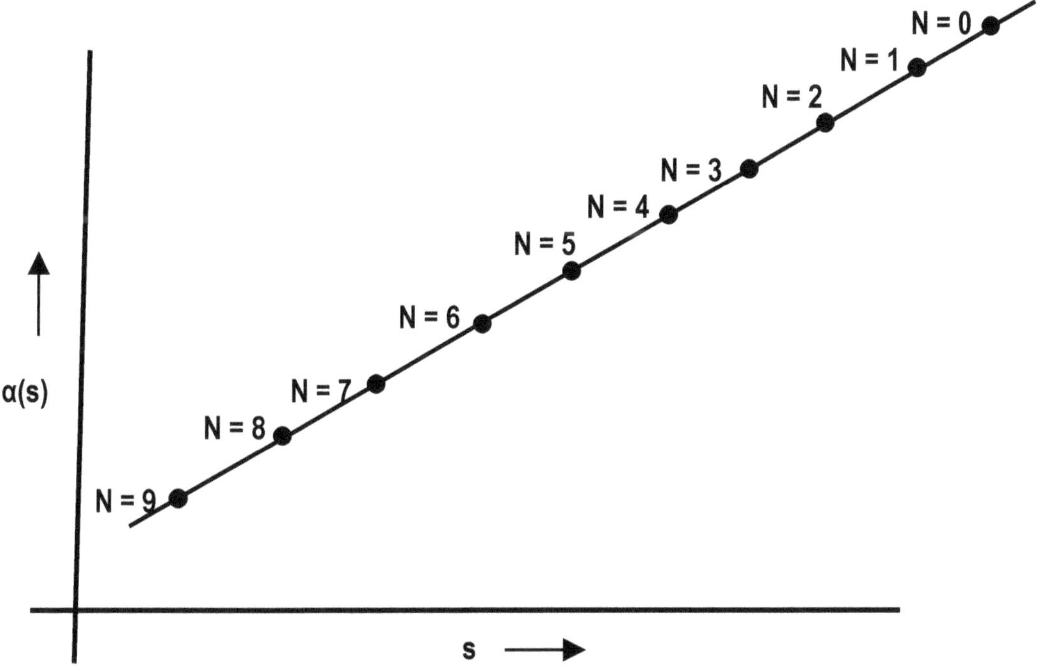

Figure 5.1. Amplitude vs. spin qualitative diagram of Blaha (2022f) where α(s) = √E = 16s + 8 in which spin is treated as a continuous parameter. Each "dot" corresponds to a HyperCosmos space.

The continuous line of dimensions corresponds to the continuous line of interpolated spins of the Regge trajectory since the dimension total of a space d_{dN} satisfies[18]

$$d_{dN} = 16(4s + 2)^2 \qquad\qquad (5.1)$$

where s is the spin of the lowest spin fermion in the space. The continuity of s implies the continuity of the value of the d_{dN} dimensions.[19]

We thus see that the concept of a continuous line of dimensions in the ProtoCosmos makes sense. It also "explains" the 1/x self-binding potential in the Dirac equation. The 1/x form is the only potential form that does not introduce [mass] dimensions in the Dirac equation for dimensions. As an additional benefit it leads to a scaling energy eigenvalue, thus to a geometric scaling d_{dN}, and thence to agreement with the derivation of d_{dN} from a consideration of the form of creation/annihilation operators in fermion wave functions.

The solution of the Dirac equation based on the dimension line view in the ProtoCosmos is a Majorana form of wave function with no implied charge (or other quantum numbers). Again this is consistent with the dimension physical interpretation.

The role of the Dirac equation is solely to create clumps of discrete dimensions.

Note: Since the dimensions line is *continuous,* clumps of dimensions can be extracted arbitrarily from its dimension continuum, and grouped to form HyperCosmos spaces with discrete sets of dimensions.

---

[18] See Blaha (2022f) eq. 1.8.

[19] The continuity in the total dimensions d_{dN} also implies that the discrete HyperCosmos spaces can be extended to a continuous set of spaces introducing the possibility of transitions between spaces.

6. Cosmos Evolution

6.1 The UniDimension Dimension Space

In chapters 1 and 4 we showed that a one dimension "UniDimension" space of continuous dimension may have a self-potential force that cause an aggregate of discrete dimensions to form for each energy E_N. Each aggregate is a space of a finite number of discrete dimensions. Together they constitute the HyperCosmos spectrum as visualized in Fig. 6.1..

$-d_{dN}$	Space-time Dimension r	Blaha Space Number N = O$_s$
0	$-\infty$	∞
$-1/16^2$	-12	15
$-1/8^2$	-10	14
$-\tfrac{1}{4}^2$	-8	13
$-\tfrac{1}{2}^2$	-6	12
-1^2	-4	11
-2^2	-2	10
-1024^2	16	1
-2048^2	18	0

Figure 6.1. The energy levels $E_N = -d_{dN}$ aggregates of the UniDimension ProtoCosmos Model (not drawn to scale). The various energy level dimension aggregates match the HyperCosmos spaces spectrum. The energy levels are symbolically depicted on a continuous dimensions line.

6.2 Aggregate Generation by a Unification Space Transformation

In chapter 3 we showed each aggregate space can be viewed as the result of a General Relativistic transformation from one dimension in the FRF of the space's unification space.

Thus the dimension aggregation of each space by the dimension self-force is alternately viewable as the result of a General Relativistic transformation in the space's unification space. A one dimension FRF is the seed from which each HyperCosmos space is generated.

6.3 Cosmic Evolution

The ProtoCosmos continuous dimension space becomes through its self-force the set of HyperCosmos spaces. At all stages one has only one primordial substance – dimension. The particles and interactions, which are dimension-based, only appear as mathematical "embellishments" when universes are instantiated (created).

6.4 Dimension as the Primordial Substance

For over 2500 years philosophers have speculated on the existence of a primordial substance or element. Initially they considered the possibility of earth, air, fire and water, or some combination thereof, might be the primordial substance. Recently the concept has been reduced to the fundamental particles of Nature: quarks, leptons, and bosons – rather a large number of primordial substances. .

This book pushes the search for a primordial substance to a new candidate: dimension. Starting from a ProtoCosmos consisting of a continuous line of dimensions subject to a 1/x agglutination force we obtain discrete aggregations of dimensions that constitute the spaces of the HyperCosmos.

Thus there is a progression from the ProtoCosmos' continuous line of dimensions to the HyperCosmos spaces with discrete dimensions, to the universes of particles and internal symmetry groups embodying dimensions. These quantities within universes consist of dimensions encapsulated in mathematics trappings.

The dimension aggregates support the instantiation (creation) of universes. Within the universes particles and interactions emerge. In the preceding series of books we have shown:

1. Each Fundamental Fermion corresponds to a "dressed" single dimension.

2. Each Internal Symmetry irreducible representation consists of a set of complex dimensions implicit in the structure of the space's dimension array. The symmetry results from the algebra associated with the irreducible representation. The consequent interactions (forces) are artifacts of local gauge considerations.

3. The Higgs bosons also emerge from dimension and space-time considerations See Blaha (2022f).

4. General Relativity, alone and in combination with internal symmetries, follows from the transformations in the Unification space of each HyperCosmos space.

5. The groups and interactions (the trappings) follow directly from the structure of the HyperCosmos space dimension arrays and the requirement of a canonical formulation.

Everything Physical embodies primordial dimensions embellished with a direct mathematical formulation.

6.5 Generation of Everything from One Dimension

The HyperUnification Space formalism developed in chapters 3 and 4 shows that the entire HyperCosmos can be "derived" from one dimension element in the HyperUnification Space.

If one is to think of a Creation from a primordial origin, the concept of Creation from a single primordial dimension element presents itself as the simplest. This origin does not appear to allow a deeper formulation. On this basis it appears that Physics now has a strong basis.

6.6 The Purpose of Physics

The purpose of Physics, in the view of some, is to establish mathematical relations between all physical phenomena. In the author's view the intellectual-conceptual elements that give understanding in human terms are missing in this approach.

The author believes the purpose of Physics is to establish the most simple, ultimate physical framework (the origin) from which all Physical phenomena is consequent. The rationale for this view is also simple: the philosophically hypothesized Prime Mover, the origin of Everything, necessarily begins Creation from this origin.

The author's simple dimension-based theory accomplishes this purpose to the greatest extent that is possible within the framework of our current understanding of Physical phenomena. It is the deepest, and most complete, explanation of the Cosmos from a Physics perspective.

The historical trend of Physical primordial elements is:

Fire-Earth-Air-Water $\rightarrow$ Atoms $\rightarrow$ Quarks, Leptons, ... $\rightarrow$ Space $\rightarrow$ Dimension

REFERENCES

Akhiezer, N. I., Frink, A. H. (tr), 1962, *The Calculus of Variations* (Blaisdell Publishing, New York, 1962).

Bjorken, J. D., Drell, S. D., 1964, *Relativistic Quantum Mechanics* (McGraw-Hill, New York, 1965).

Bjorken, J. D., Drell, S. D., 1965, *Relativistic Quantum Fields* (McGraw-Hill, New York, 1965).

Blaha, S., 1995, *C++ for Professional Programming* (International Thomson Publishing, Boston, 1995).

_______, 1998, *Cosmos and Consciousness* (Pingree-Hill Publishing, Auburn, NH, 1998 and 2002).

_______, 2002, *A Finite Unified Quantum Field Theory of the Elementary Particle Standard Model and Quantum Gravity Based on New Quantum Dimensions™ & a New Paradigm in the Calculus of Variations* (Pingree-Hill Publishing, Auburn, NH, 2002).

_______, 2004, *Quantum Big Bang Cosmology: Complex Space-time General Relativity, Quantum Coordinates™ Dodecahedral Universe, Inflation, and New Spin 0, ½, 1 & 2 Tachyons & Imagyons* (Pingree-Hill Publishing, Auburn, NH, 2004).

_______, 2005a, *Quantum Theory of the Third Kind: A New Type of Divergence-free Quantum Field Theory Supporting a Unified Standard Model of Elementary Particles and Quantum Gravity based on a New Method in the Calculus of Variations* (Pingree-Hill Publishing, Auburn, NH, 2005).

_______, 2005b, *The Metatheory of Physics Theories, and the Theory of Everything as a Quantum Computer Language* (Pingree-Hill Publishing, Auburn, NH, 2005).

_______, 2005c, *The Equivalence of Elementary Particle Theories and Computer Languages: Quantum Computers, Turing Machines, Standard Model, Superstring Theory, and a Proof that Gödel's Theorem Implies Nature Must Be Quantum* (Pingree-Hill Publishing, Auburn, NH, 2005).

_______, 2006a, *The Foundation of the Forces of Nature* (Pingree-Hill Publishing, Auburn, NH, 2006).

_______, 2006b, *A Derivation of ElectroWeak Theory based on an Extension of Special Relativity; Black Hole Tachyons; & Tachyons of Any Spin.* (Pingree-Hill Publishing, Auburn, NH, 2006).

_______, 2007a, *Physics Beyond the Light Barrier: The Source of Parity Violation, Tachyons, and A Derivation of Standard Model Features* (Pingree-Hill Publishing, Auburn, NH, 2007).

_______, 2007b, *The Origin of the Standard Model: The Genesis of Four Quark and Lepton Species, Parity Violation, the ElectroWeak Sector, Color SU(3), Three Visible Generations of Fermions, and One Generation of Dark Matter with Dark Energy* (Pingree-Hill Publishing, Auburn, NH, 2007).

_______, 2008a, *A Direct Derivation of the Form of the Standard Model From GL(16) (Pingree-Hill Publishing, Auburn, NH, 2008).*

_______, 2008b, *A Complete Derivation of the Form of the Standard Model With a New Method to Generate Particle Masses Second Edition* (Pingree-Hill Publishing, Auburn, NH, 2008)

______, 2009, *The Algebra of Thought & Reality: The Mathematical Basis for Plato's Theory of Ideas, and Reality Extended to Include A Priori Observers and Space-Time Second Edition* (Pingree-Hill Publishing, Auburn, NH, 2009).

______, 2010a, *Operator Metaphysics: A New Metaphysics Based on a New Operator Logic and a New Quantum Operator Logic that Lead to a Mathematical Basis for Plato's Theory of Ideas and Reality* (Pingree-Hill Publishing, Auburn, NH, 2010).

______, 2010b, *The Standard Model's Form Derived from Operator Logic, Superluminal Transformations and GL(16)* (Pingree-Hill Publishing, Auburn, NH, 2010).

______, 2010c, *SuperCivilizations: Civilizations as Superorganisms* (McMann-Fisher Publishing, Auburn, NH, 2010).

______, 2011a, *21st Century Natural Philosophy Of Ultimate Physical Reality* (McMann-Fisher Publishing, Auburn, NH, 2011).

______, 2011b, *All the Universe! Faster Than Light Tachyon Quark Starships & Particle Accelerators with the LHC as a Prototype Starship Drive Scientific Edition* (Pingree-Hill Publishing, Auburn, NH, 2011).

______, 2011c, *From Asynchronous Logic to The Standard Model to Superflight to the Stars* (Blaha Research, Auburn, NH, 2011).

______, 2012a, *From Asynchronous Logic to The Standard Model to Superflight to the Stars volume 2: Superluminal CP and CPT, U(4) Complex General Relativity and The Standard Model, Complex Vierbein General Relativity, Kinetic Theory, Thermodynamics* (Blaha Research, Auburn, NH, 2012).

______, 2012b, *Standard Model Symmetries, And Four And Sixteen Dimension Complex Relativity; The Origin Of Higgs Mass Terms* (Blaha Reasearch, Auburn, NH, 2012).

______, 2013a, *Multi-Stage Space Guns, Micro-Pulse Nuclear Rockets, and Faster-Than-Light Quark-Gluon Ion Drive Starships* (Blaha Research, Auburn, NH, 2013).

______, 2013b, *The Bridge to Dark Matter; A New Sibling Universe; Dark Energy; Inflatons; Quantum Big Bang; Superluminal Physics; An Extended Standard Model Based on Geometry* (Blaha Reasearch, Auburn, NH, 2013).

______, 2014a, *Universes and Megaverses: From a New Standard Model to a Physical Megaverse; The Big Bang; Our Sibling Universe's Wormhole; Origin of the Cosmological Constant, Spatial Asymmetry of the Universe, and its Web of Galaxies; A Baryonic Field between Universes and Particles; Megaverse Extended Wheeler-DeWitt Equation* (Blaha Reasearch, Auburn, NH, 2014).

______, 2014b, *All the Megaverse! Starships Exploring the Endless Universes of the Cosmos Using the Baryonic Force* (Blaha Research, Auburn, NH, 2014).

______, 2014c, *All the Megaverse! II Between Megaverse Universes: Quantum Entanglement Explained by the Megaverse Coherent Baryonic Radiation Devices – PHASERs Neutron Star Megaverse Slingshot Dynamics Spiritual and UFO Events, and the Megaverse Microscopic Entry into the Megaverse* (Blaha Research, Auburn, NH, 2014).

______, 2015a, *PHYSICS IS LOGIC PAINTED ON THE VOID: Origin of Bare Masses and The Standard Model in Logic, U(4) Origin of the Generations, Normal and Dark Baryonic Forces, Dark Matter, Dark Energy, The Big Bang, Complex General Relativity, A Megaverse of Universe Particles* (Blaha Research, Auburn, NH, 2015).

______, 2015b, *PHYSICS IS LOGIC Part II: The Theory of Everything, The Megaverse Theory of Everything, U(4)⊗U(4) Grand Unified Theory (GUT), Inertial Mass = Gravitational Mass, Unified Extended Standard Model and a New Complex General Relativity with Higgs Particles, Generation Group Higgs Particles* (Blaha Research, Auburn, NH, 2015).

______, 2015c, *The Origin of Higgs ("God") Particles and the Higgs Mechanism: Physics is Logic III, Beyond Higgs – A Revamped Theory With a Local Arrow of Time, The Theory of Everything Enhanced, Why Inertial Frames are Special, Universes of the Mind* (Blaha Research, Auburn, NH, 2015).

______, 2015d, *The Origin of the Eight Coupling Constants of The Theory of Everything: U(8) Grand Unified Theory of Everything (GUTE), S^8 Coupling Constant Symmetry, Space-Time Dependent Coupling Constants, Big Bang Vacuum Coupling Constants, Physics is Logic IV* (Blaha Research, Auburn, NH, 2015).

______, 2016a, *New Types of Dark Matter, Big Bang Equipartition, and A New U(4) Symmetry in the Theory of Everything: Equipartition Principle for Fermions, Matter is 83.33% Dark, Penetrating the Veil of the Big Bang, Explicit QFT Quark Confinement and Charmonium, Physics is Logic V* (Blaha Research, Auburn, NH, 2016).

______, 2016b, *The Periodic Table of the 192 Quarks and Leptons in The Theory of Everything: The U(4) Layer Group, Physics is Logic VI* (Blaha Research, Auburn, NH, 2016).

______, 2016c, *New Boson Quantum Field Theory, Dark Matter Dynamics, Dark Matter Fermion Layer Mixing, Genesis of Higgs Particles, New Layer Higgs Masses, Higgs Coupling Constants, Non-Abelian Higgs Gauge Fields, Physics is Logic VII* (Blaha Research, Auburn, NH, 2016).

______, 2016d, *Unification of the Strong Interactions and Gravitation: Quark Confinement Linked to Modified Short-Distance Gravity; Physics is Logic VIII* (Blaha Research, Auburn, NH, 2016).

______, 2016e, *MoND: Unification of the Strong Interactions and Gravitation II, Quark Confinement Linked to Large-Scale Gravity, Physics is Logic IX* (Blaha Research, Auburn, NH, 2016).

______, 2016f, *CQ Mechanics: A Unification of Quantum & Classical Mechanics, Quantum/Semi-Classical Entanglement, Quantum/Classical Path Integrals, Quantum/Classical Chaos* (Blaha Research, Auburn, NH, 2016).

______, 2016g, *GEMS Unified Gravity, ElectroMagnetic and Strong Interactions: Manifest Quark Confinement, A Solution for the Proton Spin Puzzle, Modified Gravity on the Galactic Scale* (Pingree Hill Publishing, Auburn, NH, 2016).

______, 2016h, *Unification of the Seven Boson Interactions based on the Riemann-Christoffel Curvature Tensor* (Pingree Hill Publishing, Auburn, NH, 2016).

______, 2017a, *Unification of the Eleven Boson Interactions based on 'Rotations of Interactions'* (Pingree Hill Publishing, Auburn, NH, 2017).

______, 2017b, *The Origin of Fermions and Bosons, and Their Unification* (Pingree Hill Publishing, Auburn, NH, 2017).

______, 2017c, *Megaverse: The Universe of Universes* (Pingree Hill Publishing, Auburn, NH, 2017).

______, 2017d, *SuperSymmetry and the Unified SuperStandard Model* (Pingree Hill Publishing, Auburn, NH, 2017).

______, 2017e, *From Qubits to the Unified SuperStandard Model with Embedded SuperStrings: A Derivation* (Pingree Hill Publishing, Auburn, NH, 2017).

______, 2017f, *The Unified SuperStandard Model in Our Universe and the Megaverse: Quarks, ... ,* (Pingree Hill Publishing, Auburn, NH, 2017).

______, 2018a, *The Unified SuperStandard Model and the Megaverse SECOND EDITION A Deeper Theory based on a New Particle Functional Space that Explicates Quantum Entanglement Spookiness (Volume 1)* (Pingree Hill Publishing, Auburn, NH, 2018).

______, 2018b, *Cosmos Creation: The Unified SuperStandard Model, Volume 2, SECOND EDITION* (Pingree Hill Publishing, Auburn, NH, 2018).

______, 2018c, *God Theory* (Pingree Hill Publishing, Auburn, NH, 2018).

______, 2018d, *Immortal Eye: God Theory: Second Edition* (Pingree Hill Publishing, Auburn, NH, 2018).

______, 2018e, *Unification of God Theory and Unified SuperStandard Model THIRD EDITION* (Pingree Hill Publishing, Auburn, NH, 2018).

______, 2019a, *Calculation of: QED $\alpha = 1/137$, and Other Coupling Constants of the Unified SuperStandard Theory* (Pingree Hill Publishing, Auburn, NH, 2019).

______, 2019b, *Coupling Constants of the Unified SuperStandard Theory SECOND EDITION* (Pingree Hill Publishing, Auburn, NH, 2019).

______, 2019c, *New Hybrid Quantum Big_Bang–Megaverse_Driven Universe with a Finite Big Bang and an Increasing Hubble Constant* (Pingree Hill Publishing, Auburn, NH, 2019).

______, 2019d, *The Universe, The Electron and The Vacuum* (Pingree Hill Publishing, Auburn, NH, 2019).

______, 2019e, *Quantum Big Bang – Quantum Vacuum Universes (Particles)* (Pingree Hill Publishing, Auburn, NH, 2019).

______, 2019f, *The Exact QED Calculation of the Fine Structure Constant Implies ALL 4D Universes have the Same Physics/Life Prospects* (Pingree Hill Publishing, Auburn, NH, 2019).

______, 2019g, *Unified SuperStandard Theory and the SuperUniverse Model: The Foundation of Science* (Pingree Hill Publishing, Auburn, NH, 2019).

______, 2020a, *Quaternion Unified SuperStandard Theory (The QUeST) and Megaverse Octonion SuperStandard Theory (MOST)* (Pingree Hill Publishing, Auburn, NH, 2020).

______, 2020b, *United Universes Quaternion Universe - Octonion Megaverse* (Pingree Hill Publishing, Auburn, NH, 2020).

______, 2020c, *Unified SuperStandard Theories for Quaternion Universes & The Octonion Megaverse* (Pingree Hill Publishing, Auburn, NH, 2020).

______, 2020d, *The Essence of Eternity: Quaternion & Octonion SuperStandard Theories* (Pingree Hill Publishing, Auburn, NH, 2020).

______, 2020e, *The Essence of Eternity II* (Pingree Hill Publishing, Auburn, NH, 2020).

______, 2020f, *A Very Conscious Universe* (Pingree Hill Publishing, Auburn, NH, 2020).

______, 2020g, *Hypercomplex Universe* (Pingree Hill Publishing, Auburn, NH, 2020).

______, 2020h, *Beneath the Quaternion Universe* (Pingree Hill Publishing, Auburn, NH, 2020).

______, 2020i, *Why is the Universe Real? From Quaternion & Octonion to Real Coordinates* (Pingree Hill Publishing, Auburn, NH, 2020).

______, 2020j, *The Origin of Universes: of Quaternion Unified SuperStandard Theory (QUeST); and of the Octonion Megaverse (UTMOST)* (Pingree Hill Publishing, Auburn, NH, 2020).

______, 2020k, *The Seven Spaces of Creation: Octonion Cosmology* (Pingree Hill Publishing, Auburn, NH, 2020).

______, 2020l, *From Octonion Cosmology to the Unified SuperStandard Theory of Particles* (Pingree Hill Publishing, Auburn, NH, 2020).

______, 2021a, *Pioneering the Cosmos* (Pingree Hill Publishing, Auburn, NH, 2021).

______, 2021b, *Pioneering the Cosmos II* (Pingree Hill Publishing, Auburn, NH, 2021).

______, 2021c, *Beyond Octonion Cosmology* (Pingree Hill Publishing, Auburn, NH, 2021).

______, 2021d, *Universes are Particles* (Pingree Hill Publishing, Auburn, NH, 2021).

______, 2021e, *Octonion-like dna-based life, Universe expansion is decay, Emerging New Physics* (Pingree Hill Publishing, Auburn, NH, 2021).

______, 2021f, *The Science of Creation New Quantum Field Theory of Spaces* (Pingree Hill Publishing, Auburn, NH, 2021).

______, 2021g, *Quantum Space Theory With Application to Octonion Cosmology & Possibly To Fermionic Condensed Matter* (Pingree Hill Publishing, Auburn, NH, 2021).

______, 2021h, *21ˢᵗ Century Natural Philosophy of Octonion Cosmology , and Predestination, Fate, and Free Will* (Pingree Hill Publishing, Auburn, NH, 2021).

______, 2021i, *Beyond Octonion Cosmology II : Origin of the Quantum; A New Generalized Field Theory (GiFT); A Proof of the Spectrum of Universes; Atoms in Higher Universes* (Pingree Hill Publishing, Auburn, NH, 2021).

______, 2021j, *Integration of General Relativity and Quantum Theory: Octonion Cosmology, GiFT, Creation/Annihilation Spaces CASe, Reduction of Spaces to a Few Fermions and Symmetries in Fundamental Frames* (Pingree Hill Publishing, Auburn, NH, 2021).

______, 2022a, *New View of Octonion Cosmology Based on the Unification of General Relativit and Quantum Theory* (Pingree Hill Publishing, Auburn, NH, 2022).

______, 2022b, *The Gold Dust Beneath Hypercomplex Cosmology* (Pingree Hill Publishing, Auburn, NH, 2022).
qqaa
______, 2022c, *Passing Through Nature to Eternity: ProtoCosmos, HyperCosmos, Unified SuperStandard Theory* (Pingree Hill Publishing, Auburn, NH, 2022).

______, 2022d, *HyperCosmos Fractionation and Fundamental Reference Frame Based Unification: Particle Inner Space Basis of Parton and Dual Resonance Models* (Pingree Hill Publishing, Auburn, NH, 2022).

______, 2022e, *A New UniDimension ProtoCosmos and SuperString F-Theory Relation to the HyperCosmos* (Pingree Hill Publishing, Auburn, NH, 2022).

______, 2022f, *The Cosmic Panorama: ProtoCosmos, HyperCosmos, Unified SuperStandard Theory (UST) Derivation* (Pingree Hill Publishing, Auburn, NH, 2022).

Eddington, A. S., 1952, *The Mathematical Theory of Relativity* (Cambridge University Press, Cambridge, U.K., 1952).

Fant, Karl M., 2005, *Logically Determined Design: Clockless System Design With NULL Convention Logic* (John Wiley and Sons, Hoboken, NJ, 2005).

Feinberg, G. and Shapiro, R., 1980, *Life Beyond Earth: The Intelligent Earthlings Guide to Life in the Universe* (William Morrow and Company, New York, 1980).

Gelfand, I. M., Fomin, S. V., Silverman, R. A. (tr), 2000, *Calculus of Variations* (Dover Publications, Mineola, NY, 2000).

Giaquinta, M., Modica, G., Souchek, J., 1998, *Cartesian Coordinates in the Calculus of Variations* Volumes I and II (Springer-Verlag, New York, 1998).

Giaquinta, M., Hildebrandt, S., 1996, *Calculus of Variations* Volumes I and II (Springer-Verlag, New York, 1996).

Gradshteyn, I. S. and Ryzhik, I. M., 1965, *Table of Integrals, Series, and Products* (Academic Press, New York, 1965).

Heitler, W., 1954, *The Quantum Theory of Radiation* (Claendon Press, Oxford, UK, 1954).

Huang, Kerson, 1992, *Quarks, Leptons & Gauge Fields 2nd Edition* (World Scientific Publishing Company, Singapore, 1992).

Jost, J., Li-Jost, X., 1998, *Calculus of Variations* (Cambridge University Press, New York, 1998).

Kaku, Michio, 1993, *Quantum Field Theory*, (Oxford University Press, New York, 1993).

Kirk, G. S. and Raven, J. E., 1962, *The Presocratic Philosophers* (Cambridge University Press, New York, 1962).

Landau, L. D. and Lifshitz, E. M., 1987, *Fluid Mechanics 2nd Edition*, (Pergamon Press, Elmsford, NY, 1987).

Misner, C. W., Thorne, K. S., and Wheeler, J. A., 1973, *Gravitation* (W. H. Freeman, New York, 1973).

Rescher, N., 1967, *The Philosophy of Leibniz* (Prentice-Hall, Englewood Cliffs, NJ, 1967).

Rieffel, Eleanor and Polak, Wolfgang, 2014, *Quantum Computing* (MIT Press, Cambridge, MA, 2014).

Riesz, Frigyes and Sz.-Nagy, Béla, 1990, *Functional Analysis* (Dover Publications, New York, 1990).

Sagan, H., 1993, *Introduction to the Calculus of Variations* (Dover Publications, Mineola, NY, 1993).

Sakurai, J. J., 1964, *Invariance Principles and Elementary Particles* (Princeton University Press, Princeton, NJ, 1964).

Weinberg, S., 1972, *Gravitation and Cosmology* (John Wiley and Sons, New York, 1972).

Weinberg, S., 1995, *The Quantum Theory of Fields Volume I* (Cambridge University Press, New York, 1995).

INDEX